MISSION ROSETTA

Exploration of the comet
Churyumov - Gerasimenko

Magdalena Gassner

Ilias Thiesseas
6774 Tschagguns, Austria
iliasthiesseas@gmail.com

Second edition

We are not lawyers. This book and the content provided herein are simply for educational purposes. Every effort has been made to ensure that the content provided is accurate and helpful for our readers. However, this is not an exhaustive treatment of the subjects. No liability is assumed for losses or damages due to the information provided. You are responsible for your own choices, actions, and results. You should consult your attorney for your specific publishing and disclaimer questions and needs.

The original work was written in German and published under the title
Mission Rosetta: Erforschung des Kometen Tschurjumow-Gerassimenko
Translated by Ilias Thiesseas

Printed in Poland

Front cover image by Ilias Thiesseas

Book design by Ilias Thiesseas

Book ISBN: 979-8646319723

ABOUT THE
PUBLISHER

Ilias Thiesseas is a professional publishing company founded in Austria, following a clear goal: To benefit society and authors, giving valuable information to the world.

We, at Ilias Thiesseas, created a new branch in our company that supports new authors.

Note that *Mission Rosetta: Exploration of the comet Churyumov - Gerasimenko* by *Magdalena Gassner* is a part of our VWA Publikation Professional Publishing Program™, an increasingly popular program for first-time writers to get their ideas into the world. We want authors to focus on what's important to them: writing. Ilias Thiesseas deals with the publishing and also contributes to junior writers' financial independence, so they can fulfil their dreams.

ABSTRACT

Comets play an essential role in the creation of our planetary system as well as for the conditions of our life on earth. Through them water and possibly also pre-biological organic substances reached the earth. During this cosmic evolution fundamental physical and chemical processes took place. The Rosetta Mission wanted to get to the bottom of these processes. The Rosetta probe was launched on 2 March 2004 by the European Space Agency. On 12 November 2014, the landing module Philae landed on the comet Churyumov-Gerasimenko.

This book deals with physical and scientific aspects. It also describes the technical configuration of Rosetta and discusses areas that have contributed to the execution of the mission.

This book covers many areas of physics and was created with the help of literature and internet sources. The aim of this book is to describe the Rosetta Mission and to put the findings of the comet

Churyumov-Gerasimenko into a cosmological context.

CONTENT

Introduction

1. **Structure of the solar system**

2. **Kepler's laws**

3. **Comets**

   3.1   The core

   3.2   The coma

   3.3   The tail

4. **Comet 67P/Churyumov-Gerasimenko and 46P/ Wirtanen**

   4.1   Comet orbits

   4.2   Comet 67P/Churyumov-Gerasimenko

   4.3   46P/Wirtanen

5. **Configuration of the Rosetta probe and the Philae lander**

6. **Philae**

7. **Course of mission**

   7.1   Ariane 5G+

   7.2   Further mission progress

7.3   Hibernation and end of the mission

8.   Summary of the main events

9.   Findings

Conclusion

References

List of figures

Footnotes

INTRODUCTION

Comets have always been of special interest in astronomy. They suddenly appear in the sky and then disappear again inexplicably. In the Middle Ages comets were called messengers of fate or even bad luck. Natural scientists like Aristotle or Ptolemy already put up theories about the occurrence of comets. Edmond Halley is considered the first astronomer who was able to determine the orbit of such a comet in 1682.

Since comets are considered to be the origin of our solar system, the European Space Agency hopes that the Rosetta probe will finally clarify the question of the origin of the Earth and our life on Earth. In doing so, it will be necessary to go deep into the matter of astronomy. The European Space Agency decided to take up this challenge and began planning the project in 1992. Research on a comet is intended to provide information about the process of evolution.

ESA is the first space agency in the world to succeed in landing a probe on a comet. The mission was led by Andrea Accomazzo. The Rosetta Orbiter and the lander Philae are the main components of the Rosetta probe.

In this book, all the processes of the Rosetta Mission are described and its findings are reflected.

A lecture by Dr. Rainer Best in October 2016 awakened my interest in the Rosetta Mission. Dr. Rainer Best is an aerospace engineer who was one of the main people responsible for the Rosetta Mission until 2004. He was responsible for the design, the construction and for the execution of various tests on the spacecraft.

1. STRUCTURE OF THE SOLAR SYSTEM

Our solar system is around 4.5 billion years old. The galaxy, which is made up of the sun, other stars, planetary systems and gas nebulae, is called the Milky Way. The center of our solar system is the sun. It takes 27000 light years, i.e. 27000 years at the speed of light, to reach the centre of the Milky Way from our solar system. Our solar system needs 210 million years to orbit around the Milky Way. The solar system orbits it at a speed of 240 kilometres per second.

Eight planets orbit the sun and reflect its light. Because of the gravitational force, it is orbited by the planets: Mercury, Venus, Earth, Mars, Jupiter, Saturn, Uranus and Neptune. [1]

Figure 1: The approximate sizes of the planets in relation to each other.

The inner four planets closest to the Sun, Mercury, Venus, Earth and Mars, are called terrestrial planets. Further away from the Sun follow the gas planets Jupiter, Saturn, Uranus and Neptune. The inner and outer planets are separated by the asteroid belt. Beyond Uranus and Neptune follow the transneptunian objects, which also include the Kuiper Belt. Finally follows the first postulated Oort Cloud.[2]

Not only these eight named planets orbit the sun, also the moons of these planets as well as comets and asteroids orbit the sun. Apart from our solar system, there are also billions of other solar systems. As already mentioned, the sun forms the centre of our solar system. It has a diameter of 1.4 million kilometers and has a mass of 333.000 earth masses. In its interior there is a temperature of about 15 million degrees. [3]

2. KEPLER'S LAWS

The German natural philosopher, mathematician, astronomer, astrologer, optician and evangelical theologian Johannes Kepler lived from 1571 to 1630 and established three laws for planetary motion that are still valid today. These laws are not only valid for planets, but also for moons or asteroids. Kepler's laws also apply to the comet Churyumov-Gerasimenko.[4]

Kepler's First Law:

„The planets move on elliptical orbits. In one of their foci is the sun."[5]

The first Kepler's law is about the orbits on which the planets move. It follows from this law that the distance between the planets and the sun is constantly changing. The perihelion of the earth to the sun, the smallest distance, is 147.1 million kilometres. The aphelion, the point at which the distance from the earth to the sun is greatest, has a value of 152.1 million kilometres. The astronomical unit (AE) was introduced for the average distance Earth - Sun. One astronomical unit corresponds to 149.6 million kilometres.[6]

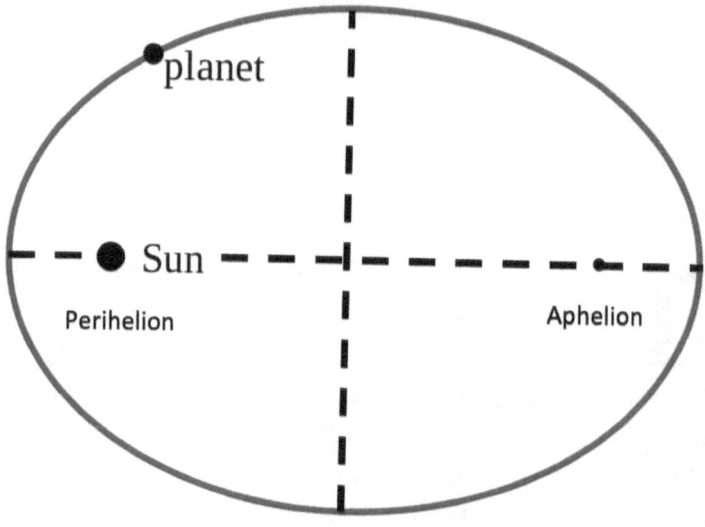

Figure 2: Kepler's first law describes the shape of the planetary orbits

Kepler's Second Law:

„The sun-planet connecting line sweeps over areas of equal size in equal times."

This line is also known as the driving beam.[7]

From Kepler's second law we can conclude that the

planets do not move at a constant speed. For the earth, in the aphelion, i.e. in the months of June and July, there is an orbital speed of about 29.3 km/s. In December and January the earth is in the perihelion, where the orbital speed is about 30.3 km/s. The closer the planet is to the sun, the faster it is.[8]

In Figure 2, the dashed lines can be imagined as a coordinate system. Then the large semi-axis corresponds to the distance from the coordinate origin to the perihelion or aphelion.

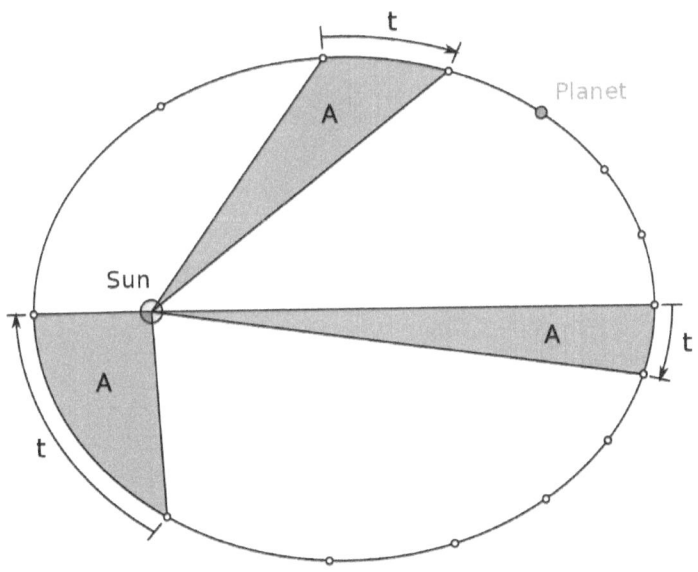

Figure 3: The sun-planet connecting line sweeps over equal areas over equal times

Kepler's Third Law:

„The squares of the orbital periods of two planets behave like the cubes (third potency) of the large semi-axes of their orbital ellipses." [9]

Formula of Kepler's third law:

$$\frac{T_1^{\,2}}{T_2^{\,2}} = \frac{a_1^{\,3}}{a_2^{\,3}}$$

T_1, T_2 = Orbit time (time for one complete orbit) of planet 1 or planet 2

a_1, a_2 = large half-axes of the orbits of planet 1 or planet 2

With the help of Kepler's third law, the connection between the large semi-axis **a** and the time required for one orbit **T** around the sun is described for a celestial body. Thus, from Kepler's third law, the large semi-axes and the remaining orbital data can be derived from the orbital period of the planets or comets around the sun.

Kepler's third law is illustrated again in Fig. 4.[10]

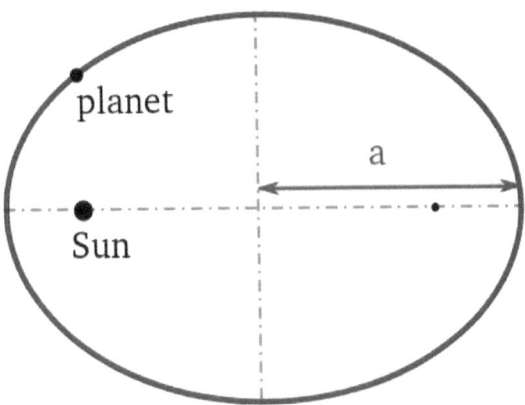

Figure 4: Orbit of a planet to illustrate Kepler's third law

3. COMETS

Comets are also known as hair or tail stars. Their diameter is about 1 to 100 kilometres.[11]

About 40% of all comets return in regular time periods. Their name is derived from their year of appearance and a letter added in alphabetical order. Comets marked with P, such as 67P/Churyumov-Gerassimenko, are periodically recurring. Throughout history they have been considered as bad luck charm, like the outbreak of plague or messengers of demons. Comets usually move around the sun.[12]

A comet consists mainly of ice and dust particles. Because of their albedo of 4%, better known as reflection radiation, they are among the darkest bodies in the solar system. Given the comet's weak albedo and the fact that it is hidden in a gas and dust envelope, it was not until 1986 that the first images of a comet's core were taken.[13]

Using the Vega-2 probe and the Giotto probe, images of the surface structure and the core were taken. The VEGA-2 probe succeeded in taking the first photographs of comet Halley on 9 March 1986. However, the images with the definitely better image quality are thanks to the Giotto probe of March 14, 1986.[14]

3.1 The core

When astronomical observers speak of pseudonucleus or the nucleus, it is not the real comet nucleus. The pseudonucleus is a bright cloud surrounding the real comet nucleus. The dust particles that are inside the cloud are separated from the surface together with the outflowing gas.[15]

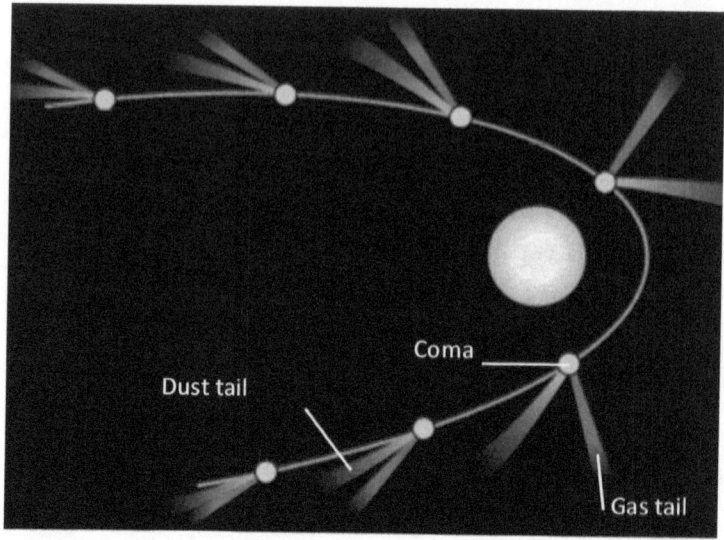

Figure 5: Representation of a comet near a sun

The structure of the real cometary core consists of solid dust particles and a mixture of loose ice types. These ice particles consist of frozen carbon monoxide, carbon dioxide, water ice, methane and other elements. The contained water ice (up to 80 % of the nucleus content) as well as parts of frozen ammonia and methane are the main elements of the nucleus. Because of this mixture they are also called icy dirt balls or dirty snowballs. The warming of the sun causes the ice mixture to change to a gaseous state. This is followed by the gas eruption, which is called jet in astronomy. As soon as a comet approaches the sun, erosion processes occur.[16]

Secondary cores are referred to as soon as they emit sufficient gas and dust through the heated surface.

It is assumed that in view of this erosion, dust particles as well as larger parts of the comet crumble away. On the one hand, such a splitting off does not occur due to any recognizable influences, on the other hand, such a process only occurs due to the action of the sun or the planets close to the orbit. The parts which are separated from the comet can be up to several meters in size. By observations nuclear fissions could already be detected. In this case the cometary core splits into two or more segments. In addition, processes have also been observed in which the nucleus was torn apart by tidal forces.[17]

3.2 The coma

If a comet is about five AU away from the sun, then there is an interaction between comet and solar wind. The solar wind, which is a stream of charged particles, comes from the sun. Here the ice particles in the nucleus are influenced by the solar radiation. A shell of water vapour, hydroxyl (a hydrogen atom binds to an oxygen atom) and carbon monoxide then forms the coma of the comet. During this process the solidified water of the comet's surface is transformed into the gaseous state of aggregation. This is called sublimation. This process affects between 10 and 15 % of the side facing the sun. In addition, the dust particles of the comet are also carried along. [18]
The diameter of the coma can even become larger

than the diameter of the sun. The coma plays a role in the research of the cometary nucleus, because it allows us to make more precise statements about the volatile components. These ice transformed into the gas phase by evaporation can be investigated by the gases escaped via sublimation. The spherical shape of the coma can be explained by the collision of the outflowing molecules. These spread around the comet in all directions in the same way. Due to the electromagnetic radiation emitted by the sun, electrons are partially removed from the gas or banished from the atomic shell. This is referred to as ionisation. The plasma tail or ion tail is then created by the interaction of the solar wind and its magnetic field with the plasma. The accelerated dust particles in the coma play an essential role in the formation of the dust tail. The result is the creation of the dust tail. Due to the light pressure pointing away from the sun and the gravitation directed towards the sun, the dust particles are stimulated to form the dust tail. Spectroscopy is important in astronomy because it enables the quantitative and qualitative investigation of gases. Karl Wurm and Polidore Swings, for example, owe their knowledge of the chemical composition of comets to spectral analysis. In 1943 they showed that comets are very reactive molecules and unstable ions. [19]

3.3 The tail

The comet's tail can extend to over 300 million kilometres. That is twice the distance from Earth to the sun. The only cometary feature visible from Earth is the tail. However, this spectacle can only be observed once or twice per decade from Earth. The brightness of a comet decreases with each passing by the sun, because this process causes a loss of mass. [20]

A distinction is made between the Type I tail, Type II tail and Type III tail. The Type I tail is also called plasma tail, gas tail or ion tail. Ionised, electrically charged nitrogen and carbon compounds are present. Each of the molecules and atoms in it contributes to the luminosity of the comet. The more than 100 million km long tail is the trademark of Type I. It runs relatively straight. Throughout its flight, the Type I tail is averted from the sun. In the Type II tail, also called the dust tail, the dust particles form the comet tail. These dust particles move away from the coma. In contrast to Type I, this type of tail has a crooked and at the same time weaker tail. However, the dust tail is also shorter. Due to the radiation pressure of the sun the dust particles leave the coma, so that the low-mass comet particles accelerate and the heavy particles hardly change their position. The type III tail is a counter-tail or anti-tail, which is very rarely seen. This tail is caused by a geometric reflection effect and does not originate from the comet itself. [21]

4. COMET 67P/ CHURYUMOV- GERASIMENKO AND 46P/ WIRTANEN

46P/Wirtanen was the Rosetta Mission's planned target comet. However, since the launch of the Rosetta probe had to be postponed due to a false launch of the Ariane 5 ECA launcher, Wirtanen no longer proved to be the more favorable target comet. Therefore the comet 67P CHU GER was chosen as a replacement target for the Rosetta mission. CHU GER is also a member of the Jupiter family, since, like the other members of this comet family, it can be detected to have an aphelion near the orbit of Jupiter. The Rosetta mission was finally launched on 2

March 2004. [22]

4.1 Comet orbits

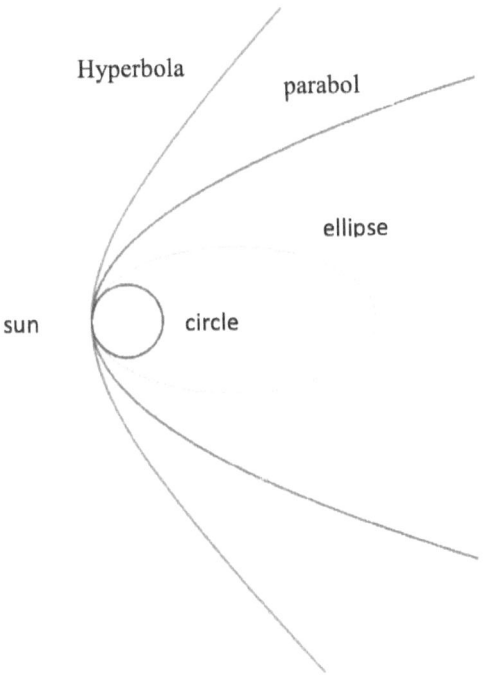

Figure 6: Possible comet orbits

If you discover a comet, only a short part of the orbit is known. One likes to use a parabola for this area for path modelling. However, orbits in the form of a parabola do not exist, because any small disturbance leads to a deviation from the parabolic path. After a longer observation a distinction is made between two different comet orbits: the ellipse or the hyperbola.[23]

Apart from the different types of comet orbits, there are also short-periodic and long-periodic comets. The short-periodic comets have their origin in the Oort cloud and the Kuiper belt. Their orbital period is not more than 200 years. A typical characteristic of short period comets is their minimal inclination against the ecliptic. The long-periodic comets are located in the Oort cloud, which surrounds the sun. However, the existence of this huge cloud is based on theoretical hypotheses. There is no definitive proof yet.

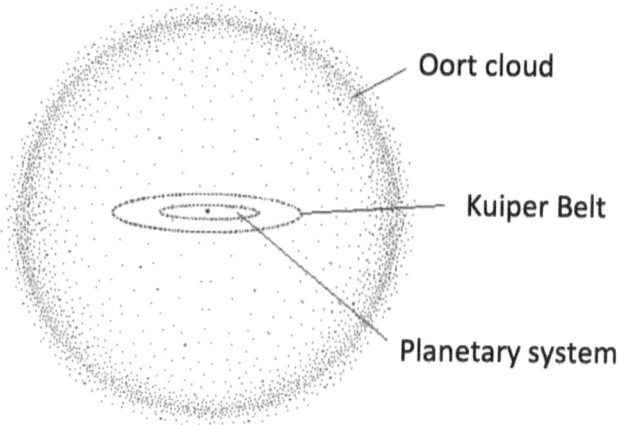

Oort cloud

Kuiper Belt

Planetary system

Figure 7: Outer region of our solar system with Kuiper belt and Oort cloud

It can also happen that comets leave our solar system forever due to planetary disturbances. These are called non-periodic. [24]

4.2 Comet 67P/Churyumov-Gerasimenko

Comet 67P/Churyumov-Gerasimenko was discovered in 1969 at the Institute of Astrophysics in Alma-Ata by Klim Ivanovich Churyumov and Svetlana Gerassimenko. Before 1840 67P was not observable because it was at least 4 AU away from the Sun. The orbit of Jupiter influences the comet so strongly that its distance in the perihelion decreases more and more. Since 1959 the perihelion distance is estimated to be 1.3 AU.

The Properties Of Comet 67P/ Churyumov-Gerasimenko:

The comet core has the dimensions of about three kilometers by five kilometers. 67P takes six years and 203 days to orbit around the sun. In the perihelion the comet is 193 million kilometers away, in the aphelion 855 million kilometers (1.289 AU or 5.717 AU). The orbit is strongly eccentric and inclined 7.1 degrees against the ecliptic. For the rotation around its own axis 67P needs about 12.6 hours. The dust production rate is about 2100 kilograms per second at 1.36 AU and about 95 kilograms per second at 1.85 AU. The gas production rate is about 10^{28} molecules per second at 1 AU.[25]

4.3 46P/Wirtanen

Comet 46P/Wirtanen is part of the Jupiter family. This comet was discovered on 18 January 1984 by Carl Alvar Wirtanen, a US-American astronomer.

The comet 46P/Wirtanen is considered to be a short-periodic comet. The orbital period of this comet is five years and 158 days. The perihelion is 1.055 AU, the aphelion 5.126 AU. The eccentricity of 0.659 is comparable to comet 67P. The orbital inclination against the ecliptic is however more pronounced with 11.764 degrees. The large semi-axis is 3.095 AU.[26]

5.

CONFIGURATION OF THE ROSETTA PROBE AND THE PHILAE LANDER

A cuboid of aluminium forms Rosetta's basic structure. This cuboid has a size of 2.8 x 2.1 x 2.0 meters. The antenna attached to the side has a length of 2.2 meters and is used for data transmission. The solar cells are also located on the side. The lander Philae is attached to the back of the probe and weighs about 100 kilograms. Philae is equipped with ten instruments, which are mounted on the top plate. At launch, the mass of the probe was three tons, half of which was hydrazine fuel (a highly toxic compound of hydrogen and nitrogen) and an oxidizer. The engines delivered a thrust of about ten Newton each. In total, about 1,670 kg of fuel was on board for the 24 dual-fuel engines [27]

Two solar panels with a length of 14 meters supply electrical energy. The energy is needed for the system and the instruments. A movable parabolic antenna with a diameter of 2.20 meters and 28 watts transmission power and additionally smaller antennas with low transmission power provide the data transmission. In this way Rosetta is controlled and the results of the experiments are transmitted to Earth.

A transmission rate of two kilobits per second was available for data transmission from Earth to the probe, while Rosetta communicated with the ground station in the range of five to 20 kilobits per second.[28]

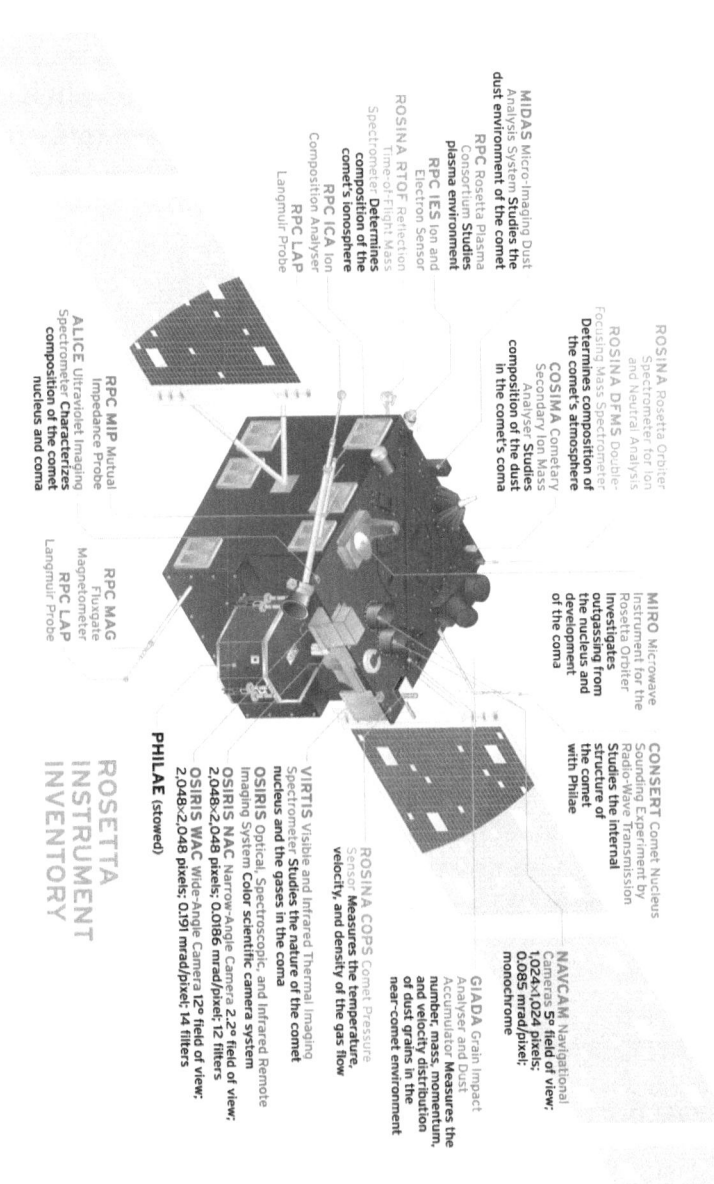

ROSINA Rosetta Orbiter Spectrometer for Ion and Neutral Analysis

ROSINA DFMS Double-Focusing Mass Spectrometer **Determines composition of the comet's atmosphere**

COSIMA Cometary Secondary Ion Mass Analyser **Studies composition of the dust in the comet's coma**

MIDAS Micro-Imaging Dust Analysis System **Studies the dust environment of the comet**

RPC Rosetta Plasma Consortium **Studies plasma environment**

RPC IES Ion and Electron Sensor

ROSINA RTOF Reflection Time-of-Flight Mass Spectrometer **Determines composition of the comet's ionosphere**

RPC ICA Ion Composition Analyser

RPC LAP Langmuir Probe

RPC MIP Mutual Impedance Probe

ALICE Ultraviolet Imaging Spectrometer **Characterizes composition of the comet nucleus and coma**

RPC MAG Fluxgate Magnetometer

RPC LAP Langmuir Probe

MIRO Microwave Instrument for the Rosetta Orbiter **Investigates outgassing from the nucleus and development of the coma**

CONSERT Comet Nucleus Sounding Experiment by Radio-Wave Transmission **Studies the internal structure of the comet with Philae**

NAVCAM Navigational Cameras 5° field of view; 1,024×1,024 pixels; 0.085 mrad/pixel; monochrome

GIADA Grain Impact Analyser and Dust Accumulator **Measures the number, mass, momentum, and velocity distribution of dust grains in the near-comet environment**

ROSINA COPS Comet Pressure Sensor **Measures the temperature, velocity, and density of the gas flow**

VIRTIS Visible and Infrared Thermal Imaging Spectrometer **Studies the nature of the comet nucleus and the gases in the coma**

OSIRIS Optical, Spectroscopic, and Infrared Remote Imaging System **Color scientific camera system**

OSIRIS NAC Narrow-Angle Camera 2.2° field of view; 2,048×2,048 pixels; 0.0186 mrad/pixel; 12 filters

OSIRIS WAC Wide-Angle Camera 12° field of view; 2,048×2,048 pixels; 0.191 mrad/pixel; 14 filters

PHILAE (stowed)

ROSETTA INSTRUMENT INVENTORY

Figure 8: Rosetta probe design

Rosetta was named after a stone dating from the Ptolemaic period (ca. 400 BC to 30 BC). In 1799 this stone was found by Napoleon's soldiers in the Nile Delta, in the town of Rosetta. Carved into the stone are demotic and ancient Greek hieroglyphics. The Rosetta Stone dates from 196 B.C. and was used to praise the Egyptian King Ptolemy. In 1822, the French scientist Champollion succeeded in deciphering the hieroglyphics with the help of the Rosetta Stone, which has been on display in the British Museum in London since 1874. As this was a breakthrough in the identification of ancient Egyptian culture, the Rosetta Mission is said to be the key to the exploration of our Earth and planetary system. [29]

Eleven scientific instruments are attached to the orbiter. The cameras and spectrometers have a wide range of functions (visual, ultraviolet, infrared, microwaves). Mass spectrometers, dust analysers and instruments for isotope analysis are used to analyse the composition of gases and dust. The radio wave experiment serves to illuminate the comet. The plasma detector investigates the exchange of particles with the solar wind.[30]

Instruments:

CONSERT:

CONSERT is used to study the comet with radio waves. The transmitter is located on the probe, the receiver on the lander. Radio waves with a frequency of 90 MHz are used. In this way the internal structure of the comet's nucleus can be examined.

COSIMA:

With the help of COSIMA, dust grains are collected, examined by mass spectrometry and thus the chemical composition is determined. The composition of the dust is thus investigated.

GIADA:

GIADA investigates the dust grains in the comet coma. Specifically, size, number and velocity are determined. The ratio of the dust percentage to the gas percentage is also determined.

MIDAS:

MIDAS is used to investigate the fine structure of the dust grains. It is a scanning probe microscope. MIDAS creates three-dimensional images of individual dust particles with high resolution.

MIRO:

It is a microwave spectrometer. The surface temperature of the nucleus is determined and the temperature, velocity and density of the molecules in the coma are determined.

OSIRIS:

This is a wide-angle camera and a telephoto camera that studies the comet's nucleus in different wavelength ranges.

ROSINA:

Here, two mass spectrometers are used to investigate the chemical composition of the coma. Besides velocity and temperature, the isotope composition is also analyzed.

RPC:

Electrons and ions are detected and the magnetic field of the comet is determined. Furthermore, the influence of the solar wind on the tail and coma is investigated.

RSI:

RSI Determines the very weak gravitational field of the comet and geological properties of the nucleus (how big is the comet, what mass does it have, is there a structure, what shape does it have?)

VIRTIS:

It is a spectrometer that measures in the visible and infrared range. Thus, the surface temperature can be determined and further information about the composition of the nucleus and coma can be obtained.[31]

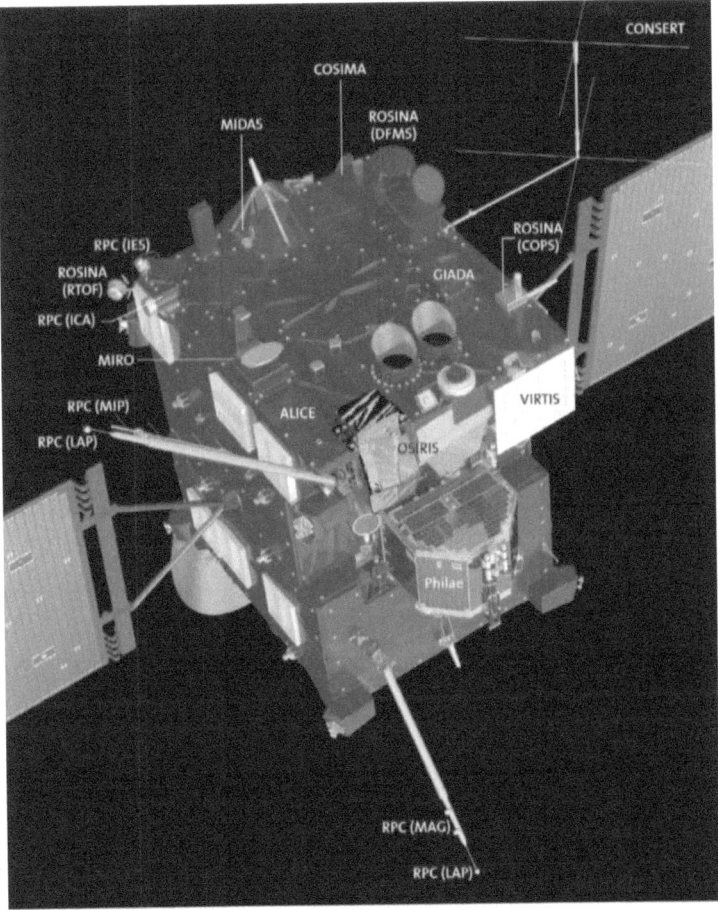

Figure 9: Rosetta probe instruments

6. PHILAE

Philae is the lander of the Rosetta probe. Philae is named after an island on the Nile near Aswan, Egypt. However, this island no longer exists, as it was flooded by the Aswan Dam. In the then existing temple an obelisk was discovered in which the names Ptolemy and Cleopatra were imprinted bilingually. This obelisk helped to decipher the ancient Egyptian characters.[32]

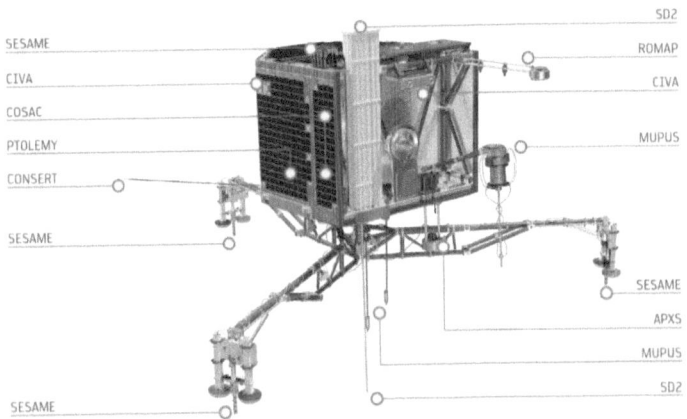

Figure 10: Instruments of Philae

Philae weighs 200 pounds. The dimensions are similar to those of a cube, namely 80 × 100 × 100 cm edge length. The structure of the lander is made of very light but stable carbon fibres, which are strengthened by composite materials. Two systems are available for the power supply. A 1000 watt-hour battery is required for the primary mission, of a duration of 65 h. For the long-term mission, a rechargeable secondary battery of 140 watt hours and a solar generator are used.

Of the six available areas of the lander, solar cells are attached to five of them. The power supply and temperature control is provided by the orbiter, from which 15 watts are used to heat the power supply.[33]

Instruments:

APXS:

APXS is a spectrometer that studies the composition directly at the comet's surface.

CIVA:

CIVA ist einerseits eine Panoramakamera, die den Landeplatz fotografieren soll. Andrerseits werden Proben untersucht, die mit einem Bohrer gewonnen wurden. Diese Untersuchung erfolgt mikroskopisch und spektroskopisch.

CONSERT:

CONSERT is used to illuminate the cometary nucleus. It is a radio wave probe. CONSERT only works in cooperation with the orbiter.

COSAC:

COSAC can examine frozen parts of the comet's surface down to a depth of 30 cm. The chemical properties of these frozen components are analysed.

MUPUS:

This instrument is used for a more precise study

of the mechanical and thermal properties of the comet. Density and strength of the material are investigated, surface temperatures and their changes are measured with several sensors. The thermal conductivity is also determined.

PTOLEMY:

PTOLEMY is used to analyse the drilling samples. It is a mass spectrometer equipped with an optional gas chromatograph.

ROLIS:

ROLIS is placed on the landing module Philae and is one of two camera systems. After landing, ROLIS takes an image of the comet's surface below the lander at a distance of 30 centimetres. The miniaturized CCD camera is located on the lander balcony. ROLIS has a resolution of 0.4 mm/pixel, which makes it possible to observe an area of 40 x 40 cm. So the size of the grains and their distribution at the landing site can be examined.

ROMAP:

ROMAP explores the magnetic field of the comet.

SD2:

SD2 is a drill with which the comet can be examined up to a depth of 30 cm.

SESAME:

SESAME is composed of three different instruments called CASSE, DIM and PP. CASSE is responsible for the research of the surface structure as well as for the determination of the material structure. With the help of DIM, measurements are carried out to determine the mass, velocity and number of particles hitting the detector. The PP instrument determines the water ice content of the comet's surface and how it changes over time.[34]

7. COURSE OF MISSION

On February 26, 2004 at 8:17 a.m., the Ariane 5G+ rocket was launched from the space center in Kourou, French Guyana.

Rosetta's actual launch date was scheduled for January 13, 2003 and her target comet was named 46P/Wirtanen. The Arianespace consortium was responsible for the space launch and a few weeks earlier tested a new model of the Ariane 5 launch vehicle. However, the rocket went off course and exploded during the test. As a result, the date was postponed to February 26, 2004.[35]

The name of the Head of Mission is Andrea Accomazzo. He is the ESA Spacecraft Operations Manager. He was born on 27 July 1970 in Domodossola,

Italy. He attended the Italian Airspace Academy as a student pilot, but later transferred to the Milan Polytechnic University. At the Polytechnic University, Andrea Accomazzo specializes in aerospace engineering. In 1999 he joined ESA (European Space Agency), where he started working on Rosetta at ESOC (European Space Operations Centre).[36]

7.1   Ariane 5G+

The Ariane rocket is a launch vehicle. The European aerospace company Airbus Group was formerly known as EADS and was considered a subsidiary of ESA. They constructed this model. Construction of the Ariane 5G+ began in 1995 and ended in 2003, with the first launch of an Ariane 5G+ taking place on 2 March 2004. A total of three Ariane 5G+ rockets were launched. The last launch of a rocket of this model took place on December 18, 2004. The reliability was 100%. There was not a single false start. The Ariane 5G+ rocket has a length of 54 m and a diameter of 5.4 m. The launch mass is 750 t and the launch thrust is 11,500 kN. 5.55 m/s^2 were achieved at the launch acceleration. [37]

Figure 11: Ariane 5G+

The difference between an Ariane 5G+ and an Ariane 5G is that the 5G+ model allows for a higher fuel load of 250 kg. In addition, longer free-flight phases were developed and Arianespace also succeeded in re-igniting the stage. As soon as the rocket is in the free flight phase, the barbecue mode is used. This mode is used to prevent a temperature difference of 200 °C. The new EPS L10, rotates with the payload around its own axis and thus enables an even distribution of solar radiation.

The Ariane 5G has two solid-fuel boosters, each consisting of three segments. The uppermost and at the same time shortest booster is decisive for the launch of the Ariane 5G, as it provides enormous thrust. The segments, which are located in the middle and at the bottom, provide thrust with the burn-up from inside to outside, because this area expands due to the burn-up. The main stage of the rocket weighs 12.5 tons and is made of very light aluminium. In order for the rocket to achieve stability, fuel or compressed gas must be filled in. When the rocket is launched, only the main engine is ignited for the first time. After seven seconds, the solid fuel boosters are activated, but unlike the main engine, they cannot be switched off after the launch.[38]

7.2 Further mission progress

Since the launch went exactly according to plan, additional fuel masses, which were intended for possible corrections of the runway, could be saved for the remaining flight. After the launch, the launch locks, also known as fastening bolts, were opened immediately. These Launch Locks were used to attach Philae to the mothership, but were no longer in use after launch. The purpose of these openings was to avoid possible thermal stress, as Philae and Rosetta were made of very different materials. Philae's structure consisted of carbon fibers, while aluminum formed the basic structure of the orbiter.

One of the most important tasks ESA had to face within the first year was commissioning. Commissioning is also known as operational testing. This process serves the purpose of checking that all equipment is in proper condition. The thermal system was the only function that was classified as faulty even before the launch. Due to these circumstances the team decided to do without this device. However, it was found that this system was functioning normally again, also because of the redundancy, after the start and during the whole flight. Whenever technical problems occurred, the ground reference models of Philae and Rosetta offered the perfect solution for the mission. The ground test models had their headquarters in Cologne, where all commands could be tested.

Roseetta started with the Swing-By manoeuvres. Rosetta had to perform this manoeuvre three times on Earth and once on Mars.

The Swing-By manoeuvre is often referred to as gravity assist, gravity manoeuvre or passing manoeuvre. A probe is brought into the gravitational field of a planet, which causes its gravitational force to deflect the probe and accelerate it. As a result, the lower-mass body (probe) absorbs energy from the higher-mass body (planet). The space probe Galileo, which was launched by NASA in 1989, also used this principle. The probe orbited the Earth once and Venus twice to gain enough momentum. Thus the probe does not reach the desired goal directly, but this concept saves propulsion costs. [39]

So whenever a space probe approaches a planet during a swing-by manoeuvre, the planet's gravity causes it to accelerate. If the probe moves away from the planet, there is a deceleration. Nevertheless, the probe gains kinetic energy during this manoeuvre. This energy comes from the planet, because it loses a little bit of orbital speed during its movement around the sun. All physical laws of conservation are fulfilled. The planet Jupiter has the largest gravitational field and is therefore a suitable candidate for a swing-by manoeuvre. A body can thus be accelerated, but can also be braked. In this case a small body moves towards a large planet, but changes its direction of flight. The speed of the small body also decreases as it gives off energy to the large body.

The acceleration manoeuvre was also used on Rosetta, because the probe with its available energy would not have made it to comet 67P. [40]

Figure 12 shows in the upper half the change of the flight path near the planet. In the lower half, the change in speed is shown schematically. Unfortunately the axis labels are missing in this representation. On the horizontal axis the time or the position of the probe might be drawn. On the vertical axis the speed of the probe is plotted.

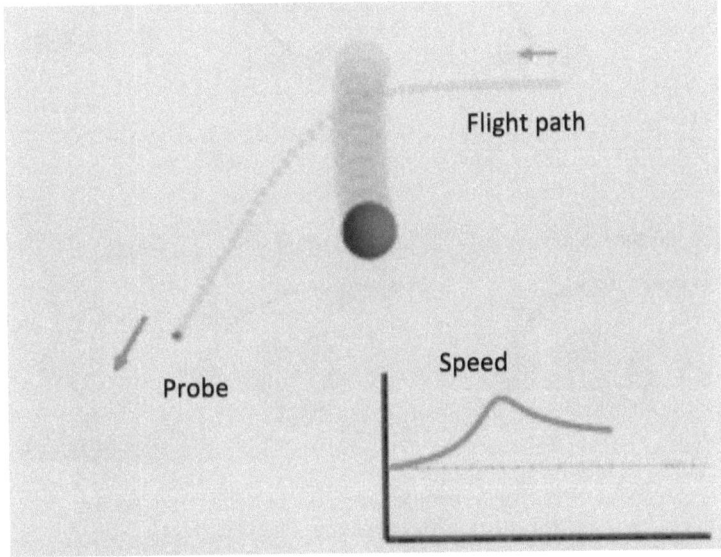

Figure 12: Acceleration of a probe through a swing-by manoeuvre

In Figure 13 below, the large blue spheres represent a planet. The smaller reddish spheres represent the space probe. The blue arrows symbolize the relative speed of the probe to the planet.

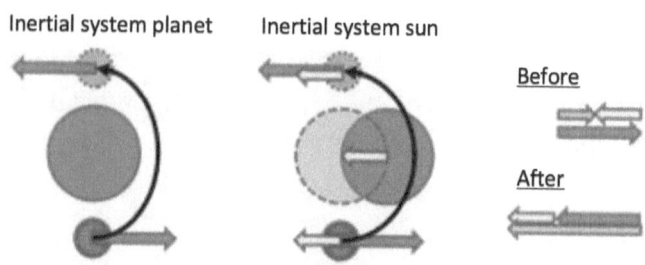

Figure 13: Swing-by manoeuvre in which a probe is accelerated. The flight direction changes by 180 degrees

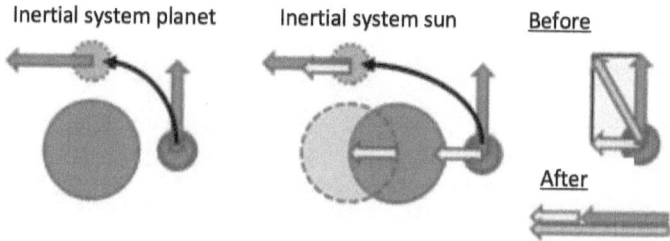

Figure 14: Swing-by manoeuvre in which a probe is deflected. The flight direction changes by 90 degrees

In Figures 13 and 14, the yellow arrows represent the speed of the planet relative to the Sun. If you now add these two vectors, you get the relative speed of the probe to the sun. This relative speed is represented by a grey arrow.

The first swing-by manoeuvre took place around the earth on 4 March 2005, with a speed change of 5.9 km/sec (21,240 km/h). During this fly-by, the probe approaches the Earth to within 1955 kilometres.

On 25 February 2007, the Mars swing-by manoeuvre was completed with a speed change of - 2.3 kilometres per second (- 8280 km/h). In contrast to the other swing-by manoeuvres, the Mars manoeuvre served to slow down the probe in order to bring it back towards Earth. The Rosetta probe approached the surface of Mars to within 250 kilometres. Because of this "small" distance, the CIVA panoramic camera was able to take an image with a resolution of 300 metres per pixel. In the background is the planet Mars, in the foreground is a small part of the probe and the back of a Rosetta solar panel.

The two subsequent manoeuvres took place around the Earth, one on 13 November 2007 with a speed change of 5.2 km/sec (18,720 km/h) and the other on 13 November 2009, with a speed change of 6.35 km/sec (22,860 km/h) in the 2009 swing-by manoeuvre.

Ultimately, the probe travelled at a total of 19.75 km/sec (71,100 km/h) using the swing-by manoeuvres.[41]

7.3 Hibernation and end of the mission

On the Mars orbit, which is one AU away from Earth, ESA was able to observe its target comet Churyumov-Gerasimenko for the first time. To bring Rosetta close to the desired comet, it was guided into Jupiter's orbit, which is five AUs away from Earth. After the probe had completed four swing-by maneuvers, it was put into hibernation at this distance. This hibernation is called Deep Space Hibernation. From this point on, the stored energy was used solely for life support of the Rosetta. This includes the receiver, the timer and heating elements of the probe.[42]

To guarantee this life support, all instruments and on-board systems were temporarily shut down. In addition, the probe went into a rotating stabilization with an orbital period of 90 seconds. Rosetta awakened as planned on January 20, 2014, and was now at a distance of 4 AU from Earth and was able to reactivate all its instruments and systems to fly to comet Churi, as the media call it.[43]

On 6 August 2014, the Rosetta probe arrived on the comet and ESA's team received the first images of its surface. ESA now had two weeks to characterise the unknown comet. On 22 August 2014, the flight engineers and those responsible for the lander Philae met at the French space agency CNES in Toulouse to find a suitable landing site for Rosetta. The distance from Rosetta to the comet was now only 100 km. From this point on, they were able to set up five options for a suitable landing site. However, the comet has a very uneven structure, which causes difficulties for the team. All five landing site options within an area of one square kilometre were analysed, none of them could meet all requirements 100 percent. The opinions of the engineers and the scientists were therefore very different.

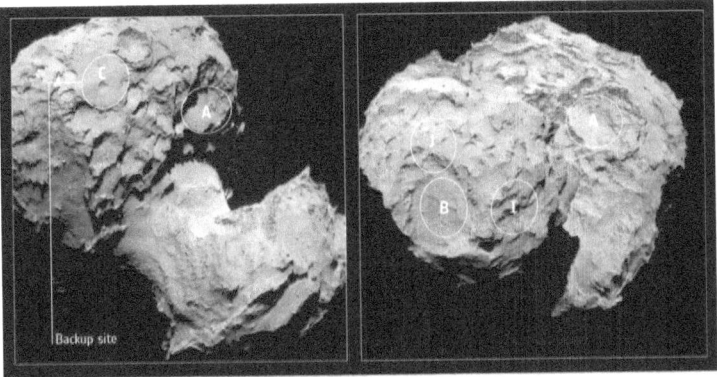

Figure 15: Comet Churyumov-Gerasimenko with its possible landing sites

In the meantime the probe was only 30 km away from the comet, which made the decision for the landing site easier. Finally the decision was made for landing site J. Here a relatively even surface had to be given to ensure a safe landing. Furthermore the day and night conditions played a decisive role. These basic conditions seemed to be most optimal at landing site J. The landing site J was christened Agilkia.

C was considered to be the secondary landing site because its light conditions seemed good. Landing site B was considered as a replacement landing site. However, it turned out that the rocks at this location would endanger the landing. The landing site A and I were not considered in round two, because they lacked fundamental conditions.[44]

Figure 16: 3D image of the landing site J on Churyumov-Gerasimenko

The probe dropped the lander Philae on the comet on November 12, 2014. From now on Rosetta served as a transmitter of information between Philae and Earth. But suddenly the team received the information that Philae's thrust system was no longer working. This thrust system was designed to press the LEM against the ground. As a result, the ESA team could now only rely on the harpoons.

At 17:03, at the exact time scheduled, ESA made history. Philae touched the ground of comet

Churyumov-Gerasimenko. However, light measurements from the solar panels indicated that Philae was in an unstable position. The Lander Module jumped away from the comet for a brief moment after impact because their harpoons had not triggered. The further flight path of the LEM was unknown. Rosetta flew to the other side of the comet at 7:00 pm as scheduled. Contact with Philae was only possible again after seven hours. The following day the signal of the Lander arrived exactly at the expected time. On November 13, 2014 the first images of Philae finally reached Earth. However, these pictures showed that Philae unfortunately landed in a crevice or on the edge of a crater. At this place the sunlight is only 1.5 hours instead of six hours. Since the solar panels were in the shade, no energy could be supplied. This made it difficult to carry out the planned research. Therefore, new methods were worked out in the Lander Control Center in Cologne in order to be able to carry out the hoped-for experiments. The team succeeded in establishing radio communication to obtain results from eight instruments of Philae.

As planned, ESA took the Philae landing module out of service on 27 July 2016 by disconnecting its batteries. Two thirds of the planned scientific experiments could be carried out.[45]

The Rosetta probe was deliberately directed at the comet on September 30, 2016 to destroy it. The Rosetta mission was now finally over.[46]

MAGDALENA GASSNER

8. SUMMARY OF THE MAIN EVENTS

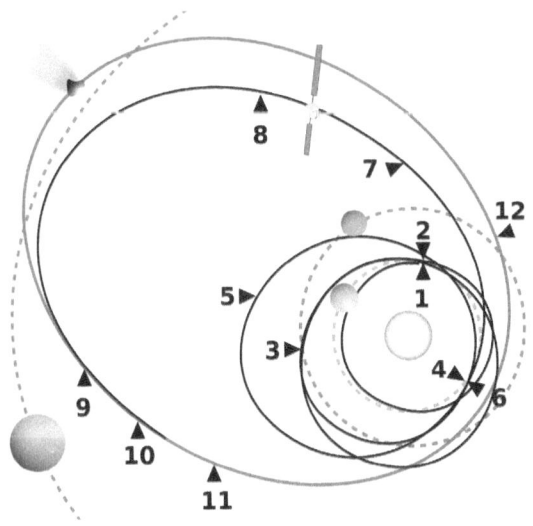

Figure 17: Rosetta's main orbital maneuvers and important events during the flight

Figure 17 shows the most important events of the mission in one picture. The orbits marked in black represent the orbits of Rosetta. The green orbits represent those of Earth, red those of Mars, brown those of Jupiter and blue those of comet 67P/Churyumov-Gerasimenko. The number 1 indicates the launch of the spacecraft and 2 stands for the first swing-by manoeuvre on Earth in March 2005. 3 indicates the Mars swing-by manoeuvre in February 2007. 4 indicates the second swing-by manoeuvre in November 2007 and 5 stands for the fly-by of the Asteoride Šteins in September 2008. The third swing-by manoeuvre on Earth took place in November 2009 and is shown in the graph with the number 6. 7 stands for the flyby of the asteroid Lutetia in July 2010 and 8 for the time when Rosetta will hibernate in June 2011. 9 stands for the awakening of the probe in January 2014 and 10 for the probe's transport towards the comet orbit. In November 2014 Philae will land on the comet, which can be found in the diagram under number 11. 12 stands for the planned end of the mission in December 2015, but the actual end of the mission took place in September 2016.

9. FINDINGS

ESA will continue to work for several years on the complete analysis of the data provided by the Rosetta probe. The results announced so far are finally described in the following paragraphs:

COMPONENTS

The main component of a comet is the dust with about 75%. The remaining 25% consists of highly porous ice.[47]

SUBSTANCES:

Apart from organic compounds such as amines, nitriles and alcohols, other nitrogen-containing molecules such as acetone, propanal and acetamide have also been found.[48]

In addition, it has been concluded that contained minerals have been formed by the intensity of solar energy.

COMA:

From the investigations an inhomogeneous composition of the coma can be concluded. Apart from the already researched molecules such as dust, water and carbon dioxide, amino acids have been added to the coma as a new component.

SUNLIGHT:

The sun is the most influential factor on the comet. It causes gas and dust eruptions which can last up to several hours. When the comet is in close proximity to the sun, sublimation of ice and the extraction of dust particles on the comet occurs. The dry ice (water, carbon monoxide and carbon dioxide) plays a decisive role in the process of the formation of the comet tail and the coma.

SURFACE STRUCTURE:

The nature of the comet resembles a chessboard. These so-called polygons are created due to the temperature differences, which causes the surface structure of the comet to break up. Terraces, craters, holes as well as crevices are created by the influence of sublimation. The gravitation there is so low that comet chunks of a few meters in diameter can be separated.

COMET CHURYUMOV-GERASIMENKO:

The porosity that still exists today is due to the creation of the comet, which was formed at a temperature below - 235 degrees Celsius.

COLLISIONS:

The observations confirm: if it is a highly porous comet, like Churyumov-Gerasimenko, then it has not suffered strong collisions. However, in other comets strong collisions have occurred due to the change of place, from the planets Uranus and Neptune.

CORRELATION BETWEEN COMET AND EARTH:

Rosetta confirms the previously existing theory that there are inorganic substances in the ice and dust of a comet. However, she came across the realization that many organic molecules, such as carbon and hydrocarbon compounds, are also found in comets and that these may have contributed to life on Earth. Unfortunately, Rosetta cannot confirm the hypothesis that comets brought water to Earth because of the ratio of hydrogen isotopes in the ice. Nor can it be proven that comets are responsible for life on Earth due to the presence of amino acids. The most important findings drawn from the Rosetta Mission's research show that the formation of complex molecules is possible on other celestial bodies and that life could therefore exist on other celestial bodies. [49]

CONCLUSION

In science it is assumed that the formation of comets in the first million years took place at low temperatures. Thus, the basic building blocks of our solar system are in them. For this reason, the Rosetta probe should help to gain more precise knowledge about the formation of the Earth by means of measurements. The European Space Agency launched the Rosetta mission on 2 March 2004, and the mission went partly as planned, but there were changes in plan in some areas, such as the first target comet.

The scientific experiments were performed by the Philae module. The Rosetta Orbiter served as a means of transport and through this module the data transfer was accomplished.

On 12 November 2014, ESA made history with its landing on the comet Churyumov-Gerassimenko. It is the first space agency to succeed in carrying out a mission of this kind.

The empirical research carried out on comet

Churyumov-Gerasimenko shows, among other things, that the main component is dust and that the sun is the cause of dust and gas eruptions. The most significant result is the presence of complex molecules on other planets, which as a result suggests the possibility of life on other cosmic bodies.

The knowledge gained will thus make a significant contribution to the exploration of further viable spaces in the universe.

MAKE A DIFFERENCE

Thank you for reading Magdalena Gassner's book! One of our missions at Ilias Thiesseas is to give young authors like Magdalena Gassner a voice. Often, junior writers do not get the chance to publish their works, because they are considered too young and inexperienced.

Many people don't know how much work goes into getting our ideas out there. We made the experience that the time to write the actual content takes about 25 percent of the whole publishing process. We also realised that one simple thing hindered new authors from getting their book out there: experience.

We, at Ilias Thiesseas, also fought hard and made some sacrifices to realise our dream: creating a publishing company which benefits both writers and society. We put our blood and souls into it, and as years went by, we got one yet simple thing: experience.

It was the lack of experience that usually made young authors give up their dreams not the lack of great ideas. They gave up their dreams of presenting society with innovative ideas, of being a writer, of changing the world.

Therefore, we created a new branch in our company that supports new authors. VWA Publikation Professional Publishing Program™ is an increasingly popular program for first-time writers to get their ideas into the world. We want authors to focus on what's important to them: writing. Ilias Thiesseas deals with the publishing and also contributes to junior writers' financial independence, so they can fulfil their dreams.

Yes, it took a lot of effort. But after seeing writers come into our offices smiling, we knew that it was worth the months of intense hard work!

We, at Ilias Thiesseas, only have a limited amount of resources and would, therefore, really appreciate your support by leaving a review on this book. Your words really make a difference in keeping dreams of new authors alive.

Thank you for your support!

REFERENCES

Web sources

46P/Wirtanen - Wikipedia. (2019). Retrieved 10. December 2019 from https://de.wikipedia.org/wiki/46P/Wirtanen

    Title: 46/Wirtanen

Kometen - Abenteuer-Universum. (2007). Retrieved 12. December 2019 from https://abenteuer-universum.de/planeten/kometen.html

    Title: Benennung

Kometen - Abenteuer-Universum. (2007). Retrieved 12. December 2019 from https://abenteuer-universum.de/planeten/kometen.html

    Title: Der Kern

Kometen - Abenteuer-Universum. (2007). Retrieved 12. December 2019 from https://abenteuer-universum.de/planeten/

kometen.html

Title: Die Koma

Ariane (Rakete) - Wikipedia. (2019). Retrieved 16. January 2020 from https://de.wikipedia.org/wiki/Ariane_(Rakete)

Title: Ariane Rakete

Der Rosetta Lander Philae - Bernd-Leitenberger. (2014). Retrieved 16. January 2020 from https://bernd-leitenberger.de/philae.shtml

Title: Der Lander

DLR. (2011). Retrieved 18. November 2019 from https://wwwdlr.de/dlr/desktopdefault.aspx/tabid-10724/1281_read9509/#/gallery/13463

Title: Instrumente Philae

DLR. (2014). Retrieved 12. August 2019 from https://www.dlr.de/dlr/desktopdefault.aspx/tabid-10907/1629_read-18794/#/gallery/23874

Title: Aufbau der Rosetta Sonde

DLR. (2014). Retrieved 20. January 2020 from https://www.dlr.de/dlr/Portaldata/1/Resources/documents/2016/Rosetta-Ergebnisse.pdf

Title: Erkenntnis über die Struktur des Kometen

ESA - *Rosetta: Wahl fällt auf Landeplatz J.* (2014). Retrieved 16. January 2020 from http://www.esa.int/Space_in_Member_States/Germany/Rosetta_Wahl_faellt_auf_L andeplatz_J

Title: Wahl fällt auf Landeplatz J

ESA - *Winterschlaf für Rosetta.* (2011). Retrieved 16. January 2020 from https://www.esa.int/Space_in_Member_States/Germany/Winterschlaf_fuer_Rosetta

Title: Winterschlaf für Rosetta

ESA: *Spacecraft Operations Manager: An interview with Andrea Accomazo.* (2004). Retrieved 9. December 2019 from https://www.esa.int/Enabling_Support/Operations/Spacecraft_Operations_Manager _An_interview_with_Andrea_Accomazzo

Title: An interview with Andrea Accomazo

Hirschler, J. (2019). *Weltall: Kometen - Weltall - Natur - Planet Wissen.* Retrieved 12. December 2019 from https://www.planetwissen.de/natur/weltall/kometen/index.html

Title: Das Geheimnis des Schweifes

Inc.,W. F. (2018). *Wikipedia.* Retrieved 16. September   2019 from https://de.wikipedia.org/wiki/Philae_%28Sonde%29#Namensgebung

Title: Philae Namensgebung

Johannes Kepler - Wikipedia. (2020). Retrieved 12. December 2019 from https://de.wikipedia.org/wiki/Johannes_Kepler

Title: Johannes Kepler

Keplersche Gesetze - Wikipedia. (2020). Retrieved 3. February 2020 from https://de.wikipedia.org/wiki/Keplersche_Gesetze

Title: Erstes Keplersches Gesetz

Keplersche Gesetze in Physik - Lernhelfer. (2015). Retrieved 27. January 2020 from https://www.lernhelfer.de/schueler-lexikon/physik/artikel/keplersche-gesetze

Title: Zweites Keplersches Gesetz

Keplersche Gesetze in Physik - Lernhelfer. (2015). Retrieved 28. November 2019 from https://www.lernhelfer.de/schueler-lexikon/physik/artikel/keplersche-gesetze

Title: Drittes Keplersches Gesetz

Komet - Wikipedia. (2018). Retrieved 11. December 2019   from https://de.wikipedia.org/wiki/Komet#Kometenbahnen

Title: Kometenbahnen

Komet - Wikipedia. (2018). Retrieved 12. December 2019 from https://de.wikipedia.org/wiki/Komet#Kometenbahnen

Title: Kometenbahnen

Kometen - Sternwarte-Eberfing. (2017). Retrieved 11. December 2019 from http://www.sternwarte-eberfing.de/Fuehrung/Objekbeschreibung/Kometen.htm

Title: Kurzperiodische- und langperiodische Kometen

Kührt, D. E. (2016). *DLR.* Retrieved 08. September 2019 from https://www.dlr.de/dlr/desktopdefault.aspx/tabid-10724/1281_read9509/#/gallery/13463

Title: Instrumente der Sonde

Kührt, D. E. (2016). *DLR.* Retrieved 09. September 2019 from https://www.dlr.de/dlr/desktopdefault.aspx/tabid-10723/1280_read9477/#/gallery/13634

Title: Rosetta-Stein und Instrumente der Sonde

Organische Verbindungen auf Komet Chury nachgewiesen. (2015). Retrieved 16.November 2019 from https://www.laborpraxis.vogel.de/or-

ganische-verbindungenauf-komet-chury-
nachgewiesen-a-499543/

Title: Organische Verbindungen auf Komet
Chury nachgewiesen

Philae (Sonde) - Wikipedia.   (2019). Retrieved
9. December 2019 from https://de.wiki-
pedia.org/wiki/Philae_%28Sonde%29#Na-
mensgebung

Title: Namensgebung Philae

Philae (Sonde) - Wikipedia. (2019). Retrieved 22. November 2019 from https://de.wikipedia.org/
wiki/Philae_%28Sonde%29#Namensgebung

Title: Namensgebung Philae

Physik Libre. (2018). Retrieved 28. January 2020
from https://physikbuch.schule/keplerslaws-
of-planetary-motion.html

Title: Zweites Keplersches Gesetz (Flächen-
satz)

Rosetta (Sonde) - Deacademic. (2010). Retrieved 14.
December 2019 from https://deaca-
demic.com/dic.nsf/dewiki/1199641

Title: Aufbau Rosetta Sonde

Rosetta (Sonde) - Wikipedia. (2020). Retrieved 16.
January 2020 from https://de.wikipedia.org/
wiki/Rosetta_(Sonde)#Winterschlaf

Title: Winterschlaf Rosetta Sonde

Sonnensystem - Wikipedia. (2018). Retrieved 28. November 2019 from https://de.wikipedia.org/wiki/Sonnensystem

Title: Sonnensystem

Swing By - Astrokiste. (2007). Retrieved 20. December 2019 from https://astrokramkiste.de/swing-by

Title: Swing by

Swing-By - Wikipedia. (2019). Retrieved 16. January 2020 from https://de.wikipedia.org/wiki/Swing-by#Prinzip

Title: Swing-by

Unser Sonnensystem - Entstehung, Aufbau & Objekte - Sternenforscher. (2019). Retrieved 28. November 2019 from https://www.sternenforscher.de/sonnensystem/

Title: Unser Sonnensystem - Entstehung, Aufbau & Objekte

Unser Sonnensystem: Planeten im Überblick - Geolino. (2017). Retrieved 15. 01 2020 from https://www.geo.de/geolino/forschung-und-technik/4917-rtkl-weltraum-unsersonnensystem

Title: Unser Sonnensystem

Komet - Wikipedia. (2011). Retrieved 12. August 2019 from https://de.dirwikipedia.org/wiki/Komet

Title: Kometen

Komet - Wikipedia. (2011). Retrieved 13. August 2019 from https://de.dirwikipedia.org/wiki/Komet

Title: Kometen

Wikipedia (2019) Retrieved 12. August 2019 from https://de.wikipedia.org/wiki/Philae_%28Sonde%29 abgerufen

Title: Philae

Wikipedia (2020) Retrieved 15 January 2020 from https://de.wikipedia.org/wiki/Rosetta_(Sonde)#Abwurf_des_Landers_auf_den_Kometen

Title: Abwurf des Landers auf den Kometen

Literaturquellen

Herrmann, J. (1993). Bertelsmann Lexikon Astronomie. Gütersloh: Bertelsmann Lexikon Verlag.

Feuerbacher, P. B. (2016). Mission Rosetta. München: GeraMond Verlag GmbH.

Möhlmann, D., & Ulamec, P. (2014). Raumsonde Rosetta - Die abenteuerliche Reise zum unbekannten Kometen. Stuttgart: Frackh-Kosmos Verlags-GmbH & Co. KG.

LIST OF FIGURES

Figure 1: The approximate sizes of the planets in relation to each other. ComputerHotline: *Solar system Scale-2*. This image is a public domain. URL: https://commons.wikimedia.org/wiki/File:Solar_system_scale-2.jpg

(last accessed: 09.05.2020)

Figure 2: Kepler's first law describes the shape of the planetary orbits. This image is a derivative of: Arpad Horvath. *Kepler's first law*. Licensed under CC-BY-SA 3.0, URL: https://commons.wikimedia.org/wiki/File:Kepler-first-law.svg

(last accessed: 14.05.2020)

Figure 3: The sun-planet connecting line sweeps over equal areas over equal times. MikeRun. *Second law of Kepler*. Licensed under CC-BY-SA 4.0, URL: https://commons.wikimedia.org/wiki/File:Second_law_of_Kepler.svg

(last accessed: 14.05.2020)

Figure 4: Orbit of a planet to illustrate Kepler's third law. 老陳. *Kepler's third law*. Licensed under CC-BY-SA 4.0, URL: https://commons.wiki-media.org/wiki/File:Kepler_third_law.svg

(last accessed: 14.05.2020)

Figure 5: Representation of a comet near a sun . This image is a derivative of: NASA. *Co-mentDiagram*. Licensed as a public domain. URL: https://commons.wikimedia.org/wiki/File:CometDiagram.png

(last accessed: 14.05.2020)

Figure 6: Possible comet orbits. URL: http://www.sternwarte-eberfing.de/Fuehrung/Ob-jekbeschreibung/Kometen.html

(last accessed: 20.01.2020)

Figure 7: Outer region of our solar system with Kuiper belt and Oort cloud. This image is a derivative of: Herbye. *Oortschewolke*. Licensed under CC-BY-SA 3.0, URL: https://com-mons.wikimedia.org/wiki/File:Oortschewolke.jpg

(last accessed: 14.05.2020)

Figure 8: Rosetta probe design. Emily Lakdawalla, Charles H. Braden, Loren A. Roberts. *Rosetta Instrument Inventory*. Licensed under CC-BY 3.0,

URL: https://commons.wikimedia.org/wiki/
File:Rosetta_Instrument_Inventory.png

(last accessed: 14.05.2020)

Figure 9: Rosetta probe instruments. Magazine from Spektrum der Wissenschaft Kompakt: Rosetta Rendezvous mit einem Komet

(last accessed: 20.01.2020)

Figure 10: Instruments of Philae. URL: https://
www.dlr.de/dlr/Portaldata/1/Resources/
bilder/missionen/rosetta/16_9/Philae_in-
struments.jpg

(last accessed: 20.01.2020)

Figure 11: Ariane 5G+. DLR German Aerospace Center. *Ariane 5ES with ATV4 on its way to ELA-3*. Licensed under CC-BY 2.0, URL: https://
commons.wikimedia.org/wiki/
File:Ariane_5ES_with_ATV_4_on_its_
way_to_ELA-3.jpg
(last accessed: 14.05.2020)

Figure 12: Acceleration of a probe through a swing-by manoeuvre. This image is a derivative of: Y tambe. *Swingby acc anim*. Licensed under CC-BY-SA 3.0, URL: https://commons.wiki-
media.org/wiki/File:Swingby_acc_anim.gif

(last accessed: 14.05.2020)

Figure 13: Swing-by manoeuvre in which a probe is accelerated. The flight direction changes by 180 degrees. Wasserkäfer. *Swingbyvxy*. Licensed under CC-BY-SA 4.0, URL: https://upload.wikimedia.org/wikipedia/commons/b/be/Swingbyvxy.png

(last accessed: 14.05.2020)

Figure 14: Swing-by manoeuvre in which a probe is deflected. The flight direction changes by 90 degrees. Wasserkäfer. *Swingbyvxy*. Licensed under CC-BY-SA 4.0, URL: https://upload.wikimedia.org/wikipedia/commons/b/be/Swingbyvxy.png

(last accessed: 14.05.2020)

Figure 15: Comet Churyumov-Gerasimenko with its possible landing sites. ESA/Rosetta/MP. *Philiae Candidate Landing Sites*. URL: http://www.esa.int/Space_in_Member_States/Germany/Rosetta_Wahl_faellt_auf_Landeplatz_J

(last accessed: 14.05.2020)

Figure 16: 3D image of the landing site J on Churyumov-Gerasimenko. ESA/Rosetta/MPS. *Philiae's primary landing site*. URL: http://www.esa.int/Space_in_

Member_States/Germany/Rosetta_Wahl_fae-llt_auf_Landeplatz_J

(last accessed: 14.05.2020)

Figure 17: Rosetta's main orbital man-euvers and important events dur-ing the flight. Kulandru mor. *Trajectoire-Rosetta.* Licensed as a public domain. URL: https://de.wikipedia.org/wiki/Roset-ta_(Sonde)#/media/Datei:Trajectoire-Roset-ta.svg

(last accessed: 20.01.2020)

FOOTNOTES

[1] Cf. Unser Sonnensystem - Entstehung, Aufbau & Objekte - Sternenforscher, 2019

[2] Cf. Sonnensystem - Wikipedia, 2018

[3] Cf. Unser Sonnensystem: Planeten im Überblick - Geolino, 2017

[4] Cf. Johannes Kepler - Wikipedia, 2020

[5] Cf. Keplersche Gsetze - Wikipedia, 2020

[6] Cf. Keplersche Gesetze in Physik - Lernhelfer, 2015

[7] Cf. Keplersche Gesetze in Physik - Lernhelfer, 2015

[8] Cf. Physik Libre, 2018

[9] Cf. Keplersche Gsetze - Wikipedia, 2020

[10] Cf. Keplersche Gesetze in Physik - Lernhelfer, 2015

[11] Cf. Herrmann, Bertelsmann Lexikon Astronomie, 1993, p. 176

[12] Cf. Kometen - Abenteuer-Universum, 2007

[13] Cf. Komet - Wikipedia, 2018

[14] Cf. Möhlmann & Ulamec, Raumsonde Rosetta - Die abenteuerliche Reise zum unbekannten Kometen, 2014, p. 23-24

[15] Cf. ebd., p. 22

[16] Cf. Kometen - Abenteuer-Universum, 2007

[17] Cf. Möhlmann & Ulamec, Raumsonde Rosetta - Die abenteuerliche Reise zum unbekannten Kometen, 2014, S. 23-24

[18] Cf. Kometen - Abenteuer-Universum, 2007

[19] Cf. Möhlmann & Ulamec, Raumsonde Rosetta - Die abenteuerliche Reise zum unbekannten Kometen, 2014, p. 45

[20] Cf. Hirschler, 2019

[21] Cf. Komet - Wikipedia, 2018

[22] Cf. Komet - Wikipedia, 2018

[23] Cf. Komet - Wikipedia, 2018

[24] Cf. Kometen - Sternwarte-Eberfing, 2017

[25] Cf. Möhlmann & Ulamec, Raumsonde Rosetta - Die abenteuerliche Reise zum unbekannten Kometen, 2014, p. 132

[26] Cf. 46P/Wirtanen - Wikipedia, 2019

[27] Cf. Rosetta (Sonde) - Deacademic, 2010

[28] Cf. DLR, 2014

[29] Cf. Philae (Sonde) - Wikipedia, 2019

[30] Cf. Kührt, DLR, 2016

[31] Vgl Kührt, DLR, 2016

[32] Cf. Philae (Sonde) - Wikipedia, 2019

[33] Cf. Der Rosetta Lander Philae - Bernd-Leitenberger, 2014

[34] Cf. DLR, 2011

[35] Cf. Möhlmann & Ulamec, Raumsonde Rosetta - Die abenteuerliche Reise zum unbekannten Kometen, 2014, p. 87

[36] Cf. ESA: Spacecraft Operations Manager: An interview with Andrea Accomazo, 2004

[37] Cf. Ariane (Rakete) - Wikipedia, 2019

[38] Cf. Rosetta (Sonde) - Deacademic, 2010

[39] Cf. Swing-By - Wikipedia, 2019

[40] Cf. Swing By - Astrokiste, 2007

[41] Cf. Möhlmann & Ulamec, Raumsonde Rosetta - Die abenteuerliche Reise zum unbekannten Kometen, 2014, p. 134

[42] Cf. ESA - Winterschlaf für Rosetta, 2011

[43] Cf. Rosetta (Sonde) - Wikipedia, 2020

[44] Cf. ESA - Rosetta: Wahl fällt auf Landeplatz J, 2014

[45] Cf. Rosetta (Sonde) - Wikipedia, 2020

[46] Cf. Philae (Sonde) - Wikipedia, 2019

[47] Cf. DLR, 2014

[48] Cf. Organische Verbindungen auf Komet Chury nachgewiesen, 2015

[49] Cf. DLR, 2014